forest phoenix

How a great forest recovers after wildfire

David Lindenmayer, David Blair, Lachlan McBurney and Sam Banks

CSIRO PUBLISHING

Dedication

In memory of those friends and colleagues who did not survive, and to the speedy recovery of all those people who did.

For the people of Marysville.

4
3
1

4
3
1

Previous pages: Burnt Alpine Ash trees in the Snobs Creek region 5 months (page iv) and 14 months (page v) after the 2009 fires.

National Library of Australia Cataloguing-in-Publication entry

Forest phoenix : how a great forest recovers after wildfire / by
David Lindenmayer ... [et al.].
9780643100343 (pbk.)
Forest regeneration – Victoria.
Forest biodiversity – Effect of fires on – Victoria.
Forest fires – Environmental aspects – Victoria.
Forest ecology – Victoria.
Lindenmayer, David.
577.342409945

Published by
CSIRO PUBLISHING
150 Oxford Street (PO Box 1139)
Collingwood VIC 3066
Australia
Telephone: +61 3 9662 7666
Local call: 1300 788 000 (Australia only)
Fax: +61 3 9662 7555
Email: publishing.sales@csiro.au
Web site: www.publish.csiro.au

Photographs are by David Blair unless otherwise noted.

Cover and text design by James Kelly
Typeset by James Kelly
Printed in China by 1010 Printing International Ltd

CSIRO PUBLISHING publishes and distributes scientific, technical and health science books, magazines and journals from Australia to a worldwide audience and cor ducts these activities autonomously from the research activities of the Commonwealth Scientific and Industrial Research Organisation (CSIRO).
The views expressed in this publication are those of the author(s) and do not necessarily represent those of, and should not be attributed to, the publisher or CSIRO.

The paper this book is printed on is certified by the Forest Stewardship Council (FSC) © 1996 FSC A.C. The FSC promotes environmentally responsible, socially beneficial and economically viable management of the world's forests.

FSC
Mixed Sources
Product group from well-managed forests and other controlled sources
www.fsc.org Cert no. SGS-COC-003963
© 1996 Forest Stewardship Council

Contents

Preface ix

Acknowledgments xiii

CHAPTER 1: A forest burning xiv

CHAPTER 2: After the fire – a forest recovers 28

CHAPTER 3: After the fire – the wildlife recovers 58

CHAPTER 4: Forest futures 86

List of scientific names 111

Further reading 113

The Eastern Yellow Robin takes advantage of gaps in the forest to catch insects and other invertebrates. This bird is clearly a persister, and has probably benefited from the removal of vegetation cover that would otherwise shelter its prey.

Preface

The February 2009 wildfires were devastating for Victorians and Australians – the worst in the nation's history in terms of loss of life and damage to property. Never before has a wildfire caused so much personal suffering in our country. Although the fires were a tragedy for so many, including some very close to us, this book does not concentrate on the human toll of the fires. Instead, it offers hope for a way forward by telling a story of the spectacular recovery which is underway in the forests. This story shows the resilience of our native plants and animals, providing us with inspiration for human recovery. This book provides a means to re-establish our connection with the forests which, contrary to the first impressions of annihilation, are rising from the ashes.

The environmental impacts of the February 2009 wildfires were profound and many spectacular stands of forest were burnt. Some of these we have worked in for more than a quarter of a century. Countless people have asked us over the past year: Will the forests ever come back? What has happened to all the animals? How many kinds of plants and animals have been lost forever? Will anything regrow? These questions were encouraged by the media, which described the forests as 'razed', 'erased' and 'wiped out'. But the forests were none of these things. It is a common misconception that *everything* is burnt in large wildfires and that nothing remains afterwards. Looking out across heavily burnt forest, it is easy to see what is missing – green canopies and thick understories. Yet critical environmental infrastructure does remain, providing a basis for further recovery. For example, while eucalypts may look devastated and many do die, the majority remain standing and will provide important habitat for wildlife in the future. This book documents the start of this ecological recovery process.

Because the recovery story is such a visual one, we have told it primarily using photographs. However, we have also tried to explain the science behind the many complex post-fire recovery strategies used by plants and animals – a subject area that is generally poorly understood. Better knowledge of fire ecology and understanding of how forests respond after fire matters. This is because fire is a key ecological process in almost all Australian landscapes. Improved understanding is also essential because of the long-term consequences of how forests are managed after fire.

Most of this book is about the wet ash or montane ash forests of the Central Highlands of Victoria. These are forests dominated by Mountain Ash, Alpine Ash and Shining Gum trees. Occasionally we also touch on forests adjacent to stands of montane ash forest, such as those dominated by Messmate, Red and Brown Stringybark, and Mountain Grey Gum. There are good reasons for our focus on ash forests – we have worked as scientific researchers in them for more than 27 years and we hope to continue working in them for at least a further two decades. We know the montane ash forests better than any other kind of forest and we believe the scientific foundations for what we describe in this book are solid.

Although this book focuses on ecological recovery following the 2009 wildfires, the science we briefly outline is equally relevant to previous fires in montane ash forests in 1851, 1926, 1932, 1939 and 1983. Similarly, many of the plant and animal recovery strategies we describe are common in many other forest and non-forest environments elsewhere in Australia, and indeed around the world.

The structure of this book

This book comprises four chapters, each with a short essay, based primarily on scientific insights, followed by a series of images (mostly photographed by David Blair) and accompanying captions that interpret the key points of the photograph.

Chapter 1 provides some ecological background to wildfire as a form of natural disturbance, including some commentary about how forest landscapes were burned in the February 2009 fire. Chapter 2 explores the extraordinary range of recovery strategies employed by native plants including trees, understorey plants and ferns. Chapter 3 focuses on the responses of animals to wildfire. Finally, Chapter 4 covers a range of issues associated with the future of the forest, and how forest and fire management can either promote or impair the ecological recovery process. We also touch on other issues such as the risks posed by climate change, the impacts of traditional clearfell logging, and the effects of post-fire salvage logging.

An important caveat – the meaning of 'recovery'

Throughout this book we have repeatedly used the word 'recovery' as well as the term 'ecological recovery'. This implies regaining something that has been lost. Recovery is undoubtedly an apt term for many species in many places, but it is not an appropriate description for all of them. Some species cannot be described as having 'recovered' because they were not 'lost' after the fire. Instead, they were either unaffected or actually prospered in the newly burned conditions. The 'invasion' of many of our long-term field sites by the Flame Robin is one of several examples. To best communicate the forests' story, we have not used an alternative to the word 'recovery', but we hope our readers will be aware of the subtle but important nuances in its meaning.

David Lindenmayer, David Blair, Lachlan McBurney and Sam Banks
April 2010

Keppel Falls near Marysville, before and after the 2009 fires.

Acknowledgments

We thank the people of Marysville. The town was our research base from 1983 until 2009.

We are deeply indebted to our long-term field team colleagues for their outstanding work in the Victorian forests: Mason Crane, Chris MacGregor, Rebecca Montague-Drake and Damian Michael.

Our work in the wet forests of Victoria has been supported by many people and organisations over the past 27 years, including: the Australian Research Council; the Victorian Department of Sustainability and Environment; Parks Victoria; VicForests; the Federal Government Department of the Environment, Water, Heritage and the Arts (including through the Commonwealth Environmental Research Facility [CERF]); the Thomas Foundation; the Hermon Slade Foundation; the Kendall Foundation; Earthwatch; and private donations from Jim Atkinson and Di Stockbridge. We also have been supported by literally thousands of volunteers over the years who have assisted us in our extensive field programs.

Over many years, we benefitted from the wonderful hospitality of John and Briarley Malcolm at Nanda Binya Lodge in Marysville – a beautiful place that was lost in February 2009. We hope that many of the special places that used to characterise Marysville will rise from the ashes – just as the forest is now doing.

Many of the insights in this short book are based on past collaborative research with outstanding colleagues, including the veritable statistical wizards Ross Cunningham, Jeff Wood, Alan Welsh and Emma Knight, and ecologists Don Driscoll, Jean Dubach, Joern Fischer, Jerry Franklin, Phil Gibbons, Malcolm Gill, Heather Keith, Gene Likens, Brendan Mackey, Mike McCarthy, Henry Nix and Hugh Possingham.

We thank Rachel Muntz, Merridee Bailey and Clive Hilliker for their wonderful assistance in many aspects of writing this book. John Stein compiled the fire severity map.

This book was championed by John Manger, publisher extraordinaire and great supporter of our work over many years.

1 | A forest burning

The forest was extraordinarily dry on the day before Black Saturday in 2009, and it was eerily still and quiet where we were working at Cambarville, about 20 kilometres from Marysville. Trampling a branch on the forest floor produced a crack that carried far into the distance – a complete contrast to the deafening firestorm that tore through Cambarville the next day.

We have worked at Cambarville since 1983 and although the area was an initial focal point, our work has since expanded across the Victorian Central Highlands region (Figure 1.1). Our research in the region has focused primarily on montane ash forest. The region covers approximately half a degree of latitude and one degree of longitude (37°20′-37° 55′S and 145° 30′-146° 20′E) – an area of about 400 000 hectares. The major towns include Healesville, Warburton, Marysville, Alexandra, Powelltown, Toolangi and Noojee. These forests are dominated by stands of Mountain Ash, Alpine Ash or Shining Gum. Mountain Ash typically occurs at altitudes between 200 and 1100 metres, Alpine Ash between 900 and 1450 metres and Shining Gum between 600 and 1300 metres.

Rainfall and fire

Rainfall data and fire are intimately linked. Back in 1983, montane ash forests were a very tough place to work. Annual rainfall exceeded 2 metres in some places. These prolonged periods of rain created a damp that produced mould on walls, disintegrated sleeping bags, and even rotted the circuitry of old-time laptop computers. Many night-time surveys for animals were abandoned due to wet weather – perhaps one night in two was lost. At least that was the experience for the first decade. That seemed to change in the mid-1990s. The forest became an easier place to work – fewer leeches, and not as much sliding on mossy logs or recently fallen wet branches. Hiking boots replaced gumboots. Meteorological data since then tell us that the annual rainfall over the past decade has been about 20% less than average, causing the forest, its plants and its animals considerable and prolonged water and heat stress prior to the 2009 fires. Similar drought conditions also preceded previous major wildfires in 1939 and 1983.

A fire approaches. The intense heat of the fire generated enormous columns of smoke, forming pyrocumulus clouds and sending burning debris and ash into the sky.

Two fires in one

The February 2009 wildfires burned over 3500 square kilometres in Victoria, of which a large proportion was high severity forest crown fire. They were more extensive than the 1983 fires in the Central Highlands region, but less widespread than those in 1939. In many respects, the February 2009 wildfires can be considered not one, but two fires. The first, during the late afternoon and evening of 7 February, was the very fast moving, high severity wildfire that led to tragic results in towns including Marysville, Kinglake and Steels Creek. The second was a far slower moving, lower severity fire, which burned over the following two weeks. As we discuss further below, the two fires had profoundly different effects on the forest, which in turn has led to markedly different trajectories of ecological recovery. These differences are among an array of factors which have contributed to the variability in fire effects and the complexity of responses to wildfire.

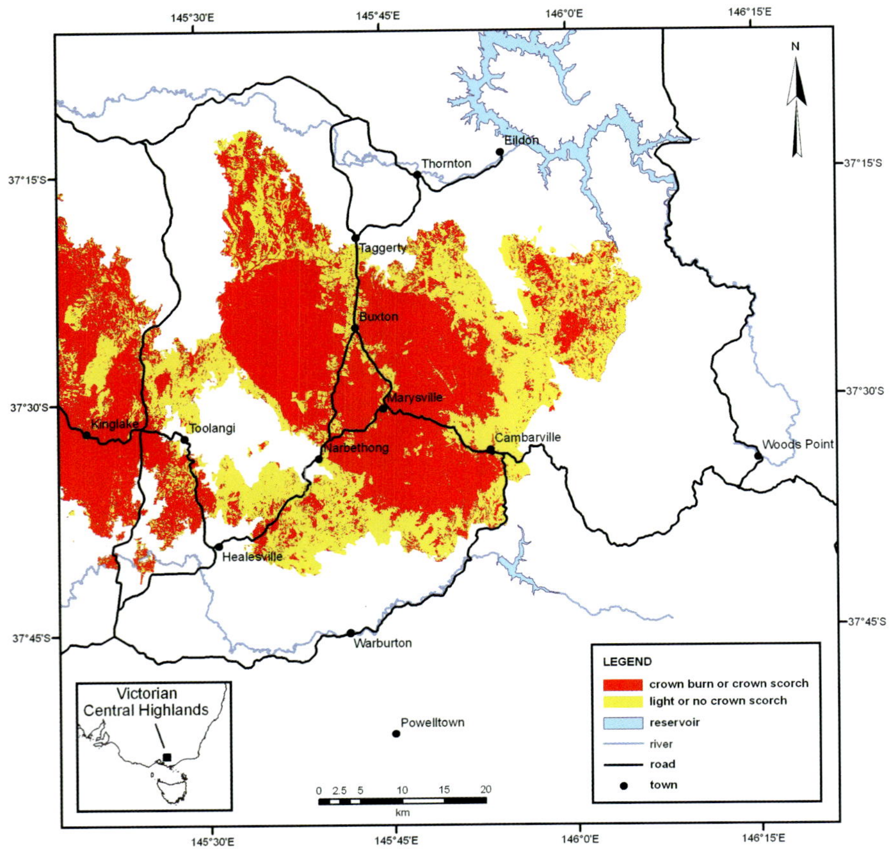

Figure 1.1 A fire severity map of the Central Highlands. This map shows the distribution of high and low severity fires in the Victorian Central Highlands, north-east of Melbourne, in February 2009. The broad location of the region that is the focus of this book is indicated on the inset in the lower left of the figure. Much of what was burnt on 7 February burnt at very high severity. The fires that continued to burn for weeks afterwards were of lower severity.

Image generated from Department of Sustainability and the Environment FireSev_09 GIS data

Fire severity

In Table 1.1 we highlight some of the many factors contributing to variation in fire severity within stands of trees. Fire can range from a 'trickle burn' in which only the ground layer or shrub layer is scorched, through to a very high severity fire where the entire canopy of very tall trees is totally consumed. The full range of severities occurred within the area burned in the 2009 wildfires – in some severely burned places even the outer layers of granite boulders were cracked and peeled away, while other places were left unburned, lush and green. These unburnt patches can be critical sources of shelter and food for animals during and after a fire. All of the factors listed in Table 1.1, except season, contributed to the variation in the severity at which the forest burned across the Central Highlands region.

Table 1.1. Some of the factors contributing to variability in wildfire severity.

Factor	Explanation/notes
Weather patterns	Temperature, wind and humidity are three of the most obvious factors contributing to fire severity and fire spread.
Timing of a fire	Fires late at night or in the morning usually burn at lower severity compared with fires during the afternoon or early evening due to lower temperatures, higher humidity and lower wind speeds at these times of day.
Seasonality of a fire	Fires in summer can have different effects on plants and animals than fires in winter or spring. For example, fires in spring may be more likely to burn nests constructed by birds or the flowers and seeds produced by some species of plants.
Landscape position	Gullies and south-facing slopes are wetter and more protected than ridges and north- or west-facing slopes. Fires are likely to be lower in severity in these areas.
Time since last fire	The amount of time elapsed since the last major disturbance influences the age of a stand of montane ash forest. A fire in an old-growth stand can have very different effects to a fire in a young forest. For example, relatively young plants may not be sufficiently mature to produce viable seeds and hence to regenerate successfully after a fire.
Severity of previous fires	A previous fire of low severity will produce different kinds of stand conditions than a previous fire of high severity.
History of logging	Forests with a recent history of logging are more fire-prone than unlogged areas.
Type of vegetation	Some kinds of forest typically burn at higher severity than others. This is related to factors like the moisture content and flammability of the vegetation and also the type, amount and vertical structure of biomass available to be burned.
Prescribed burning history*	Past prescribed burning (fuel reduction burning) may influence the severity and spread of fire in some vegetation types and in some landscapes.

*Prescribed burning has not been undertaken in montane ash forests as it is not appropriate in this forest type (see Chapter 4).

Fire regimes

Much of what we have discussed so far relates to the impact of a single fire event. However, a more important concept in fire ecology is that of the fire regime. A fire regime has four key components:

- fire severity
- fire type (e.g. crown or ground fire)
- between-fire interval (or frequency)
- season.

The typical fire regime in montane ash forests is an infrequent, high severity, stand-replacing crown fire that occurs in late summer. The forests that we see today have existed with this fire regime for thousands and perhaps even hundreds of thousands of years. Part of the area burned in the 2009 Black Saturday fires exemplified most of the key features of this fire regime. We stress in this book that plants and animals are rarely eliminated entirely from an area as a result of a single fire. Rather, it is a **series** of fires that is outside the bounds of natural variability – too many fires or too few fires over time – which can have the most significant negative impacts on biodiversity. For example, altered fire regimes currently threaten more species of birds in Australia than any other process except land clearing.

Biological legacies

Forests do not regrow from a 'blank canvas' after fire. Rather, biological legacies are the parts of a forest which survive a fire and provide a starting point for recovery. These legacies include living and dead standing trees, logs, plants, seeds, rhizomes or rootstocks, fungi, bacteria, and animals which often survive by using other legacy resources, such as logs, for shelter. Variation in the number and type of these biological legacies, together with the mosaic of burnt and unburnt forest, shapes the different ways in which a forest recovers. This variation is influenced by the life history attributes of the plants and animals themselves, such as the ability of plants to resprout, or the ability of animals to hide in burrows or hollow trees.

Our studies of ecological recovery after the 2009 fires have already produced many surprises: the extent of rocky outcrops shattered by the ferocity of the flames on Black Saturday, evidence of truly extraordinary flame heights in some areas, the depth of leaf litter consumed in some old-growth stands revealing massive root boles of ancient shrubs more than 350 years old, the pace and extent of resprouting in Messmate forests, and early invasions of spectacular birds like the Flame Robin. These surprises are a prominent part of the following chapters on plant recovery (Chapter 2) and animal population recovery (Chapter 3).

Forest giant. Mountain Ash trees are the tallest flowering plants in the world. These trees dominate the skyline of many wet forest regions in Victoria and Tasmania. Mature trees can exceed 100 metres in height and 350 years in age.

A past fire in a Mountain Ash forest. All of the eucalypts in this stand of trees regenerated after the 1939 Black Friday fires. More than 70% of the Victorian Central Highlands forests burnt in 1939. While the overstorey trees are only 70 years in age, much of the luxuriant tree fern understorey is more than 200 years old.

Forest types in the Victorian Central Highlands. The dominant forest type changes dramatically with elevation. In the Victorian Central Highlands, forests below approximately 500 metres in elevation consist of a mixture of eucalypts, including Messmate, Red Stringybark, Mountain Grey Gum and Narrow-leaved Peppermint (top left). Mountain Ash forest predominates between 500 and 1000 metres (top right), above which it gives way to Alpine Ash (bottom left). Above 1300 metres in altitude, Snow Gum woodland predominates (bottom right).

Cool temperate rainforest. This forest type commonly occurs in sheltered gullies in the Central Highlands region. It is dominated by ancient Gondwanan trees like Sassafras and Myrtle Beech (left). These are often interspersed with a dense cover of tree ferns. These forests, together with montane ash forests, are critically important for producing most of Melbourne's drinking water (right).

Understorey diversity. The wet forests of the Central Highlands support an amazing diversity of understorey and ground cover plants. Three of the more common understorey plants are the Mountain Correa, Hazel Pomaderris and Bootlace Bush (opposite). There are many species of ferns (above), mosses, liverworts and lichens, a large number of which are dependent on fallen logs on the wet forest floor (right).

A forest in flames. A tree is engulfed in the flames of the Black Saturday fires. Image: AAP Image/Andrew Brownbill

A ring of fire. A rooftop view of the 2009 fires from the eastern suburbs of Melbourne. Image: Chris Mulherin

Clouds of smoke. A satellite image showing the smoke from the fires at 3:50 pm on 7 February 2009, driven across Bass Strait by hot north-westerly winds. Image: NASA/GSFC, MODIS Rapid Response Team

Land of the long white cloud. Smoke generated from the Black Saturday fires affected areas as far away as New Zealand. Image: NASA/GSFC, MODIS Rapid Response Team

Boil your rocks off. In some areas, the severity of the fire was so extreme that the outer layers of granite boulders were fractured. Nevertheless, the trees, which initially appeared dead, began to resprout within weeks.

Underground ovens – four months on. Although the most destructive fires occurred in early February and heavy rain fell in March, in some places the fires continued to burn underground until June. The soil in these areas often had a distinctive orange colour. Once extinguished, the ash bed provided a rich area for seedlings to germinate and grow rapidly. One of our diary entries from winter 2009 reads: 'We found many "underground ovens" that were smouldering months after the fire had passed. We found one in June that was more than a metre deep and still red-hot after months of rain and winter frosts. These hidden "ovens" would collapse when you unknowingly stepped onto their fragile ceilings. Massive root systems had allowed these "sleeper" fires to continue smouldering completely hidden from the surface.'

Burnt forest near Marysville. Fire in wet forest created enormous quantities of dead stems and branches. However, even where a fire has been extremely severe, many kinds of animals make use of this dead wood. For example, several species of wood-boring beetles are attracted to dead-wood environments, where they are in turn preyed upon by other beetles, birds and mammals.

A green gully among burnt forest. Variations in landscape topography lead to different fire severity. On this slope near Cambarville (east of Marysville) the surrounding drier ridges burnt at high severity, while the wetter and more sheltered gully survived. Places like this gully are important refuges for animals and plants that can then spread back out once the surrounding habitat becomes suitable. The yellow canopies are Silver Wattle in flower.

The unpredictability of wildfire. In a high severity wildfire, even the wettest and most protected areas in the landscape can burn at high severity.

A forest destroyed? Although this is a scene of devastation, signs of new growth were appearing within weeks. Image: © The Herald and Weekly Times Photographic Collection

High severity fire. Large areas of forest were burned at very high severity, like this area on Lady Talbot Drive near Marysville. This extensive landscape is dominated by Alpine Ash and Mountain Ash trees, the vast majority of which were killed.

Fire burns both young and old trees. The 2009 wildfires burnt some of the most extensive areas of old-growth forest in mainland Australia, with trees dating from before 1800 being killed during the late afternoon and evening of 7 February 2009. Some of these trees survived several previous fires. Although the massive trees in this stand are likely to be dead, they will continue to play many important ecological roles in these forests until at least 2100.

Many areas of young, previously logged forest were burnt. All of the eucalypts in this stand of trees have been killed and they may not regenerate naturally because they were not of sufficient age to produce viable seed … if this eventuates, the resulting stand may be dominated by wattles instead.

How severe is severe? The term 'severity' has a special meaning in fire ecology. It can be measured specifically in the context of how much damage occurs in a stand of trees, and how much of the original vegetation cover is consumed by the fire. The higher the severity of the fire, the more vegetation is consumed. This image (opposite) shows the highest-severity fire, with complete shrub consumption, leaving just trunk stubs. The lack of leaves on the ground indicates that the canopy leaves have also been consumed by the fire (right).

Medium severity fire. Here, the smaller branches of shrubs are consumed and the canopy scorched, leading to extensive leaf fall. Using callipers, scientists can measure the minimum diameter of the remaining branches as one indicator of fire severity (right). In this image, scientists have used a 1-metre-square quadrat to measure the extent of leaf fall from a scorched canopy as an indicator of fire severity (bottom right).

Low severity fire. Low severity fire was typical of forests that burnt over a 2-week period following the major fire front of 7 February 2009. Areas subject to low severity fire were characterised by a green canopy and a ground level fire with scorching of the shrub layer.

Appearances can be deceiving. Most of the forest in this image has experienced a low severity burn, but it appears to be unaffected as the canopy is largely unburned.

Tide marks on tree trunks. Fire in this old-growth stand left the forest canopy intact, but removed the leaf litter and humus layer. Every hectare of old-growth forest can support tonnes of leaf litter and humus. The humus and leaf litter layers are critically important for retaining moisture and nutrients, and facilitate long-term storage of carbon in the soil. Replacing the numbered metal tags on the trees at our research sites provided a new challenge as the ground surface had often been lowered by 20–40 centimetres through the consumption of the leaf litter and humus layers, leaving the old tags too high to reach.

A clean slate? The forest does not recover from a 'blank canvas' after a bushfire. The contrast between burned logged forest and burned unlogged forest in this image highlights the 'biological legacies' that are left behind after an unlogged forest is burned. Hollows in the burned standing trees, as well as fallen logs, will provide important shelter for recovering animal populations for decades into the future. Seeds falling from the scorched forest canopy will initiate massive germination and tree recovery. The structural complexity of burned forest stands enables the persistence and recovery of the biodiversity of the forest.

2 | After the fire – a forest recovers

Montane ash forests are home to hundreds of species of plants, from giant Mountain Ash trees – the world's tallest flowering plant – to liverworts and mosses on logs on the forest floor. These different species have co-evolved with the current fire regime by employing a remarkable range of recovery strategies. We mentioned in the last chapter that the forest is not empty after a fire, and does not (and cannot) recover from a 'clean slate'. Therefore, even when it appears as if nothing living remains following a severe fire, a myriad of recovery processes are already underway, particularly where biological legacies remain. In this chapter we discuss the two broad categories of plant recovery strategies in montane ash forests: 'survivors' and 'germinators'. We also discuss plant-based biological legacies left after fire such as seeds, logs, rhizomes and rootstocks.

An ancient process. Many kinds of Australian forest have evolved in different ways with fire over time. Rare but high severity fires have occurred in this landscape over hundreds of thousands of years. What initially appears to many people to be a destroyed forest is only a brief stage in an ongoing recovery process. This image, taken a few months after the fire, shows the beginnings of regeneration on the forest floor.

The survivors

To the human eye, much of the severely burned forest looked dead after Black Saturday. However, some plants were not killed by the fire, even in the most severely burned stands of forest, and they recovered by resprouting from trunks or root balls in an incredibly short time. In montane ash forest, burnt and surviving plants use four major strategies for recovering: epicormic resprouting, general resprouting, rhizomal resprouting and lignotuberous growth. We discuss each of these plant survival strategies in more detail below.

Epicormic resprouting

A large proportion of eucalypt species have the ability to resprout, even when most of the outer layers of bark are burnt by fire. This is because they have buds along the length of strands of epicormic tissue running from the bark surface to the woody part of the tree. Some species are particularly potent epicormic resprouters like Shining Gum, Mountain Grey Gum, Messmate and Red Stringybark. A classic experimental study completed in the late 1950s found that particular species of Victorian eucalypts could have their leaves clipped off and then resprout more than 25 times before finally dying. However, other species of eucalypts, such as Mountain Ash and Alpine Ash, have very limited capacity for epicormic resprouting and are quite sensitive to wildfire. This is particularly the case for young trees, which have thin bark and are usually killed by fire.

General resprouting

Tree ferns are some of the most charismatic plants in montane ash forests and are among the first to recover after wildfire. This type of general resprouting occurs around the crown, creating verdant new foliage.

Rhizomal resprouting

Rhizomes are interconnected underground root systems that enable rapid resprouting after the above-ground part of a plant is removed by a disturbance such as fire or vegetation clearing practices like mechanical slashing. Common Bracken Fern is a classic example of a plant species that resprouts from a rhizome.

Lignotuberous growth

A lignotuber is a woody swelling at the base of some species of plants that occurs either fully or partly underground. Lignotubers contain reserves of nutrients which assist a plant to recover if its above-ground parts are destroyed. A good example of this strategy is the understorey shrub Musk Daisy Bush, which can resprout after fire. Old-growth Daisy Bushes sometimes exceed 350 years old and have recovered after several fires. The large bole of these plants is often hidden under a deep layer of leaf litter and revealed only after a fire has consumed that layer.

Resprouting plants provide some of the most spectacular scenes in the recovering forest, with the vivid greens and reds of new growth contrasting with blackened trunks. This rapid regrowth provides food and shelter for animals such as birds and possums. Of course, there are many paradoxes in any natural ecosystem and the montane ash forests are no exception. For example, tree ferns are very susceptible to drought but recover vigorously and rapidly after fire. Indeed, tree ferns may live for hundreds of years, over which time they may be subjected to many severe fires. Similarly, rainforest relics such as Myrtle Beech are strong resprouters and can recover vigorously and rapidly after fire, yet they have origins dating back to ancient Gondwanan times, more than 60 million years ago, when high severity fire was a rarity.

The seed germinators

Seed germinating plant species survive fire by very different means to survivors. If a fire is severe enough to kill a seed-germinating plant, it is also sufficiently severe to stimulate the germination of seeds already stored in the soil, or trigger the release of seed from the canopies of burnt trees.

The soil seed bank

Some species store seeds in the soil for rapid germination when conditions are favourable. Wattles are a classic example of trees that recover in this way. Wattle seeds lie dormant in the soil for many decades and possibly even hundreds of years. Germination is triggered by a fire or mechanical disturbance of the soil. This means that a particular area of forest may appear to have no living wattles at the time of disturbance but a dense layer of wattle seedlings will usually develop rapidly following a fire.

Seed dropped from a burnt canopy

After fire, some species of trees release vast quantities of seed from woody fruits in the canopy. These seeds fall to the ground and germinate. However, if a wildfire is very severe and consumes the canopy (including the fruit containing seeds), the 'seed rain' will be limited, resulting in the germination of relatively few trees. The most prominent canopy seed germinators in the montane ash forest are Mountain Ash and Alpine Ash eucalypt trees. The 'death by fire and recovery from seed' strategy leaves some species susceptible to a change in the fire regime, or the presence of other disturbances such as logging. For instance, Mountain Ash trees must reach maturity (at approximately 20 years) before they can produce viable seed. Where fires occur at shorter intervals than this, or young, logged and regenerating forests are burned, there will be few eucalypt seeds to germinate and the regenerating stand will be dominated by wattles (see Chapter 4).

Recolonising from outside the burnt area

Some plant species germinate in a burnt area from seeds that are blown in by the wind, or brought in by animals either as burs in fur or passed through the gut. This includes some native species like the Fireweeds and Bidgee Widgee, but also many invasive weeds such as Thistles, Blackberry, Ivy, Holly and Radiata Pine. These plants are early colonisers of areas where the soil layer has been disturbed either through fire or by machinery.

The extent of plant germination following a wildfire can be remarkable. We have been measuring plant species cover at 160 permanent long-term monitoring sites since 1983. Of these 160 sites, 64 were burnt in February 2009, resulting in mass seed germination. While we have counted all of the germinating plant species, the overstorey eucalypts have been particularly interesting, with estimates of up to 5 million seedlings per hectare! These seedlings compete intensely for the key resources that dictate plant survival: light, water, nutrients and space. Over the coming decades, the vast majority of seedlings will be outcompeted by their neighbours. Others will be lost through browsing by insects and mammals. This natural process of self-thinning will mean that of an initial ~5 million eucalypt seedlings germinating soon after the fire, 100–200 trees per hectare will remain after 80 years, and there will be perhaps as few as 30–40 per hectare after 200 years.

Intense competition for resources in the early stages of post-fire regeneration means that growth rates of trees are truly breathtaking. Young Mountain Ash trees can grow 1 metre per year for the first 70 years of their life, and in sheltered locations they may exceed 90 metres in final height. This illustrates the remarkable life cycle of this tree species, which commences life as a tiny seed but eventually becomes a tall forest giant. Only the coniferous Redwoods of California are taller than Mountain Ash, but it takes them four times longer to reach their top height.

Other biological legacies

Plants that have burned and died provide a source of nutrients for other living things. Indeed, many mosses, ferns and lichens grow only on fallen logs. Logs also play important structural roles, modifying or stabilising environmental conditions in a recovering stand and shielding young plants from overbrowsing by large animals such as wallabies or introduced Sambar deer. Structural biological legacies, in the form of hollow-bearing logs and standing dead and live trees, are precious sheltering sites for many forest animals. At a landscape level, patches of unburned forest are considered legacies. Plants and animals use them as critical refuges during the fire as well as a starting point for recolonisation surrounding burned forest.

Because no two fires, forests or landscapes are identical, the number, type and spatial pattern of biological legacies that occur after a fire event can vary enormously, leading to different recovery trajectories after different fires or even in different parts of a forest after the same fire. Variation in the types and numbers of biological legacies is influenced by:

- the type and severity of fire
- the age and structural diversity of the forest before it was burned
- the types of organisms that were present in that landscape before the fire, such as the kinds of resprouters and seed germinators present
- the topography of a landscape (i.e. elevation, slope, aspect)
- site characteristics such as soil type and levels of soil moisture.

The biological legacies concept is critically important because it underscores the fact that natural disturbances, including severe ones like the 2009 Black Saturday wildfires, do not 'wipe the slate clean'. Rather, what remains after a fire strongly influences the recovery of both plants and animals. This critical insight has major implications for how we manage the forests after wildfires, as we discuss in Chapter 4.

Increasing knowledge of post-fire recovery

Despite the research that has been done on natural disturbances in forests around the world, it is remarkable how much remains to be learned about their impacts, about ecological recovery, and about the best ways to manage the recovery process so as not to impair it. This is why, in mid-2009, we commenced a major set of studies on our long-term research sites in the Central Highlands of Victoria, previously established between 1983 and 1988. The main objectives of this work is to compare animal and plant populations on burnt and unburnt sites, as well as contrasting these same populations before and after the 2009 fire. This will allow us to quantify and describe the ecological values of young and slow recovering stands of forest, of which relatively little is known, and compare this to old-growth forests. Our work has already produced some unexpected new insights, but many other fascinating findings will appear only after several years' or even decades' more dedicated field research. It should not be surprising that it will take a long time for new discoveries to be made – 40 years of a researcher's working life is but a fraction of the 350-year generation time of a stand of Mountain Ash trees!

The changing forest. Steavenson Falls, near Marysville, has been a popular scenic attraction for well over a century. Over that time, the forest has been through significant changes. These changes occur on different timescales. The most rapid and dramatic change is through the passing of a wildfire. The next shortest is the seasonal change that occurs every year. The longest is the recovery of the forest from a burnt state. The blackened forest on the right was last burnt 70 years ago in 1939. It is likely that this forest looked like the right panel in 1939 before recovering spectacularly over the following decades.

A history lesson in the forest. Fires burn old forest, but also young forest. What is being burnt matters. Biological legacies from the stand before a fire are carried through into the regenerating stand. The dead trees in this old-growth forest (right) are over 70 metres tall and will remain standing for up to 80 years. These large hollow-bearing trees provide critical shelter for native animals. Soon they will be surrounded by a new cohort of regenerating young trees. This mixture of dead standing trees and young forest will provide high quality habitat for a wide range of native animals. In stark contrast, this stand of 10-year regrowth forest (left) has been highly simplified by a past clearfell logging operation followed subsequently by the 2009 fires. There are no large, hollow-bearing trees that will be 'biological legacies' at this site, and thus this stand will contain none of the hollows that are critical shelter resources for native animals for 150 years or more. Our research indicates many species of native plants have been lost from these areas.

The forest phoenix. Different species of plants have different strategies for recovery after fire. One of the most startling innovations is epicormic resprouting on the trunk. Some species of Australian plants can continue to resprout after more than 25 successive disturbances. Eucalypts display striking variation in the types of epicormic growth as shown by several tree species found in the Central Highlands region, including Shining Gum (left), Swamp Gum and Long-leaved Box (top right), Narrow-leaved Peppermint (bottom right), and Messmate (opposite).

A year in a fuzzy forest. This spectacular series of images shows the extent of epicormic development. The first image (top right) was taken a week after the 2009 fires. The second (bottom right) shows the extent of development after 3 months. After a year (opposite) the entire forest is green. This nutrient-rich growth provides critical food and shelter for native animals like the Common Ringtail Possum. A range of bird species use these areas to build their nests. This is a normal and natural process in which trees make a transition to full recovery after fire.

Regreening the landscape. Extensive areas of epicormic regrowth on some kinds of eucalypts create truly surreal but nonetheless beautiful landscapes that are also rich in native species. The dense, green young foliage growing on the trunks and large branches of recovering trees are nutrient-rich sources of food for many species of insects. These invertebrates, in turn, attract an array of species of insectivorous birds – a key part of the complex food webs which characterise natural forest ecosystems.

The orchid enigma. (Left) Most orchids are only seen at certain times of the year. Some species have a substantial underground tuber, which was an important source of food for Aboriginal people. This tuber also enables orchids to respond rapidly after fire, as in the case of this Common Bird Orchid.

Fire ferns. (Opposite) Despite their ancient origins and propensity to occur in wet gullies, tree ferns are remarkably resilient to wildfires, even very high severity ones. The large dead tree in the top centre of the photograph was killed by the 1939 fires. However, the tree ferns survived that fire and probably several before that. It is not uncommon to find 350-year-old tree ferns in montane ash forests where the overstorey eucalypts are only 70 years old.

Tree ferns were producing new fronds within months of the fire at many of our study sites.

Emerald jewels in a blackened forest. Myrtle Beech has ancient Gondwanan origins, with much of its 60-million-year evolutionary history occurring in the absence of fire. Many people, including us, have been surprised to see how vigorously this species is recovering after the 2009 fires. The size of the bole under the trunk of the huge resprouting Myrtle Beech is 2 metres in diameter and 1.5 metres in height. This demonstrates the tree's extraordinary age and hints at the number of fires it may have seen in its lifetime.

An old-growth couple: a giant tree and an old fern. The tree ferns at this site have fully recovered within a year of the fire. This image shows a large tree with a classic 'church door' fire scar at the base, indicating the area has had a history of past fire. These scars develop on the leeward side of a tree, revealing important information on the behaviour of past fires. The tree ferns in this photograph may well be the same age as the giant eucalypt (number 2) in the mid-ground.

Fire exposes a hidden history. This is an ancient Musk Daisy Bush, only about 3 metres high, but which probably germinated 100 years before Captain Cook's arrival in Australia. The bole on this shrub is more than 1 metre in diameter and was exposed by the leaf litter consumption during the 2009 fires. Since this photo was taken, this shrub has resprouted from the base. The plant has a hollow where ants have cached wattle seeds, which then germinated after the fire.

Stored energy. This Forest Lomatia has begun to recover using energy reserves in the rootstock. This is a common sight on many of our monitoring plots in the Central Highlands forests.

A rain of seeds from the burnt canopy. The state of the canopy has an important effect on the amount of germination taking place on the ground. Most people think of forest recovery beginning with a seed bank stored in the soil. However, for many Australian species such as eucalypts, the fall of seeds from the canopy after the fire is the major source of germinants. Eucalypts protect their seeds through the fire within woody fruit. In old-growth forest where the canopy was scorched but not consumed, the 'seed rain' from Mountain Ash trees was enormous. This was reflected by an extraordinary germination event, with up to 1800 eucalypt seedlings per square metre of forest floor. This is the beginning of a new generation of trees, which, if left undisturbed, will still be growing in 350 years time.

Different forest histories, different recoveries. These seedling quadrats are a part of our post-fire research. Clockwise from the top left, these show: no early seedling germination in a burnt site that had been logged just before Black Saturday; seedlings in a stand subject to low severity fire; an unburnt stand of forest; a dense carpet of fallen leaves following full scorch of the canopy from a medium severity fire – seedlings were often germinating under these leaves; snowfall which retarded the growth of germinants in high altitude sites; a massive pulse of germinants in a burnt old-growth stand in which around 1800 seedlings were counted.

Storing seeds for the big occasion. This is natural seed fall from Silver Wattle before the 2009 fire. It is estimated that the small black seeds in the pods can remain viable in the soil for as long as 200–400 years. This explains our observations that after the fire, there was massive regeneration of wattle on sites where adult wattle trees were totally absent prior to the fire.

From little things, big things grow. One of the remarkable aspects of life on Earth is that a seed smaller than a pin head can turn into the tallest flowering plant in the world. If it survives, this Mountain Ash will grow at a metre per year for the first 70 years of its life.

A fight for life. There can be literally millions of these Mountain Ash seedlings in every hectare of burned forest. Over the following years, the numbers will reduce through competition, browsing and natural thinning.

A seedling carpet. This veritable wheat field of Red Stringybark seedlings will naturally thin to a small fraction of these individuals, as shown by the mature trees in the background of this image.

Ready, set, grow! Like the eucalypts, few wattles will survive the initial race for survival, and the density of these young Mountain Hickory Wattles will reduce massively over their 60- to 80-year life span, but in that time they will produce seed which will once again be stored safely in the ground until the next fire.

Tracking a forest recovery. The germination of plants is a critical part of post-fire forest recovery. Here, field ecologist Lachlan McBurney is counting the number of seedlings of different plant species in a 1-metre-square quadrat in a logged and burnt site. This truly labour-intensive process is being repeated annually on nearly 500 plots.

Unwanted invaders. Many weeds can take advantage of recently disturbed soils, such as the Common Blackberry.

Adapted to fire. The orange colouring in this image is *Neurospora,* which is a genus of fungus that occurs around the world, and has a life cycle that is adapted to fire. The heat and chemical by-products of burning plant matter produce a sterile and nutrient-rich environment suitable for the germination and growth of these fungi. They are commonly found on recently burnt plant matter, such as trees, shrubs, grasses and even toast! Image: Geoff Cary and Malcolm Gill

A rotting lot. Fungi are an important but often overlooked part of the biodiversity of forests. Many species of fungus appeared soon after the fire in burnt stands of forest. Fungi play an important role in decomposing plant and other material, but also are a food resource for some animals such as the Bush Rat and the Mountain Brushtail Possum. Some species are found growing on a single host species, like the Beech Orange (*Cyttaria gunnii*) (right), which grows only on Myrtle Beech (opposite). Fungi are a much understudied group of organisms in all Australian environments.

Wattle seed is an important food for many species. Ants are large consumers of wattle seeds, often carrying them to their nests, where they subsequently germinate, sometimes in unexpected places.

The new and the old. The biological legacies of the pre-fire forest coupled with a regenerating new stand create a complex forest structure, and provide critical habitat for a vast range of forest-dependent animals.

3 | After the fire – the wildlife recovers

Fires can kill many individual animals. While this is tragic, ultimately it is the recovery of **populations** of animals that is most important. Animal population recovery is intimately linked with the plant response strategies discussed in the previous chapter. This is because animals need habitat to survive, which means the type and quality of vegetation that develops after a fire is critical. The reverse is also true, as plants often need animals to complete key ecological processes such as pollination and seed dispersal.

Like the plants, the species of native animals that inhabit montane ash forests have co-evolved with the present fire regime over thousands of years. The strategies they use to respond to fire are many and varied. The most common recovery strategies are: staying put, rapid new colonisation, disappearance and then relatively rapid recolonisation, and disappearance and then slow recolonisation. We briefly discuss each of these strategies below.

Survival strategies

Staying put

A surprising number of animals survive even the most severe wildfires by staying within their territories. For instance, the week before Black Saturday we fitted radio-collars to 19 Mountain Brushtail Possums in the Cambarville area, which was subsequently burned. Several months later, all 19 of these possums were found alive. Most would have survived by sheltering in the hollows of tall, large-diameter, old-growth trees. Both the diameter of these hollow trees and the possums' fur provided insulation against the extreme temperatures of the wildfire. Small mammals such as the Agile Antechinus also can survive fire by sheltering in tree hollows. Common Wombats use the same strategy at ground level, sheltering deep in their burrows. Even small numbers of animals that survive a wildfire in this way can have an important influence on post-fire population recovery – because recovery is not dependent on recolonisation from outside the fire boundary.

The Yellow-tailed Black Cockatoo is one of the largest and most mobile of any of the birds in wet eucalypt forests. Its loud and mournful cries led to one of its common names being the funeral bird. It is a genuine persisting species in these forests because it is extremely mobile and can range over large areas to find shelter and food.

Survival after a fire itself, however, is only the first hurdle for animals that use this strategy – many animals that stay in burned areas subsequently perish because of limited food and reduced number of sheltering sites. The lack of vegetation cover also leaves them more vulnerable to predators. This may be the reason why some animals actually try to spread fire – raptors have been known to drop smouldering sticks into unburned areas, as it is much easier for them to find prey in burnt areas.

Rapid new colonisation

Newly burnt forest creates conditions that are highly suitable for some species that previously did not occur in an area. These animals can quickly colonise a burned area. Surveys of birds immediately following the 2009 fire showed that the Flame Robin was one of these species. It forages for insects in open areas where ground cover plants have been burned. Recently burned forests are ideal environments for the Flame Robin, and at least one pair was recorded on every one of our long-term sites that had been disturbed by the 2009 fires.

Disappearance and then rapid recolonisation

Some animals disappear from a burnt stand of forest but then quickly recolonise, sometimes within days, weeks or months. Some species that fall into this category simply moved out of the area in advance of the fire front. The Yellow-tailed Black Cockatoo and the Crimson Rosella are good examples as these birds can easily move over large distances. On a smaller scale, mammals such as the Swamp Wallaby (or Black Wallaby) can move into wet gullies to increase their chances of surviving during a fire. Afterwards, they move back into the burnt forest as their food sources begin to regenerate.

Disappearance and then slow colonisation

Some parts of an animal's habitat can take a long time to develop. Large cavities or hollows in trees are an example. The Greater Glider requires large trees with hollows for shelter and nesting and depends on eucalypt leaves for food. Since the fires in 2009, we have found that the Greater Glider is absent from areas where the forest canopy was burnt. Furthermore, where young forest stands are burnt and there are no large hollow-bearing trees, it may be 150–200 years before suitable kinds of trees with hollows can develop and the Greater Glider can return.

The link between plant recovery and animal population recovery

Animal population recovery, whether it commences through survival or recolonisation, is intimately tied to the structure and recovery of the plants in the forest. For some animal species, the germination and growth of food plants is stimulated by fire. For example, the green carpet of young seedlings and resprouting tree ferns in montane ash forests are key food resources for surviving animals like the Mountain Brushtail Possum.

The importance of logs

Wildfire is also a critical part of the process of accruing new shelter resources for many animals. Fire produces significant quantities of dead wood, including logs, which are important shelter sites for many ground-dwelling animals. In fact, as fallen logs begin to rot, they become the basis for an entire assemblage of organisms called the saproxylic community. This community contains many fascinating invertebrates such as Velvet Worms, Springtails and Harvestmen, most of which few people would know about. These communities of saproxylic invertebrates undergo major changes in numbers as logs become increasingly decayed with time after wildfire.

The importance of hollows

In the long term, the death of, and damage to, standing trees resulting from wildfire can stimulate the development of cavities. These hollows are essential habitat for many animals. Since 1983, we have been researching the formation and collapse of hollow-bearing trees, and the dependence of montane ash forest wildlife on these trees. Over the coming decades, we plan to document the rate at which new hollows are formed in fire-damaged trees and how these trees facilitate the recovery of wildlife populations in burned forest. We already know, however, that hollows suitable for wildlife are most likely to form in large trees. Standing, burned large trees that contain hollows can make a regenerating stand suitable for Leadbeater's Possum within 10 years of a fire. However, burned young forest that does not contain these biological legacies will not provide suitable habitat for Leadbeater's Possum for 100–200 years. The prevalence of trees with hollows in a burnt stand is dependent on a number of factors: (1) the age of forest at the time it was burned – more large hollow-bearing trees occur in an old-growth stand that is burned than in young forest that is burned; (2) whether the trees were alive or dead immediately prior to a fire – dead trees are far more susceptible to being fully consumed by a fire than living trees; (3) the severity of a fire – a high severity fire will remove most large dead trees that existed in a stand at the time of the fire.

The importance of unburnt patches

For many animals, patches of green forest are critical for their survival in an otherwise burnt landscape. Like most fires, the 2009 Black Saturday fires were patchy in many places, leaving a mosaic of burned and unburned areas, as well as places subject to moderate severity fire. The Greater Glider is one species that is dependent on the presence of unburnt patches for survival, as it requires large hollow-bearing trees for shelter and green eucalypt leaves for food. We have found Greater Gliders in burned old-growth forest, but only near forest patches with an unburned canopy. Over time, dispersing animals from these unburned refugia will promote the speed of post-fire ecological recovery in surrounding burned areas.

Inter-relationships between animal populations and green refugia, as well as the presence of biological legacies, are critical factors influencing post-fire recovery. However, management practices like post-fire salvage logging can have a profound effect on ecological recovery. These forest and wildlife management issues are among the core topics of the next chapter.

Wildfires, wildlife and trees with hollows

Trees with hollows are a key resource for more than 40 species of vertebrates and countless (and largely unknown) species of invertebrates in montane ash forests. Cavity-dependent species include the nationally endangered Leadbeater's Possum as well as other iconic and high profile animals like the Yellow-bellied Glider, the Sooty Owl and the Yellow-tailed Black Cockatoo. Australia lacks animals like woodpeckers that can rapidly create tree hollows. Cavity formation is dependent on the activities of termites and fungi and is a long and slow process in the dense hard wood of Australian eucalypts – typically taking hundreds of years. Trees with cavities suitable for animals such as Leadbeater's Possum and the Greater Glider are usually at least 150 years old and sometimes twice that age. Importantly, past research has shown that the seven species of arboreal marsupials which inhabit montane ash forests typically use different kinds of trees with hollows. The Greater Glider and the Sugar Glider use tall, intact living or recently dead trees, whereas the Leadbeater's Possum and the Mountain Brushtail Possum use short, dead and highly decayed trees. These highly decayed trees have a high probability of collapse. Moreover, all of these decayed, dead trees on our burned long-term monitoring sites were all but totally consumed by the 2009 fires. This indicates there will be a shortage of decayed, dead trees in burned areas of montane ash forest in the future. Unfortunately, finding alternative accommodation for the animals that depend on these kinds of trees with hollows will not be straightforward. Another of our research projects – a 10-year nest box study – has revealed that providing human-made artificial cavities is a generally unsuccessful strategy for the Leadbeater's Possum, although some success has been achieved in other forest types. Work in North America has demonstrated that it is possible to rapidly accelerate cavity formation in conifers by 'girdling' or stripping bark from trees, and by firing fungi-infected pellets into tree trunks. There is no equivalent research in Australia on how to promote cavity development in living eucalypt trees – but the time to commence it must surely be now.

The ultimate survivor. Wombats are the world's largest burrowing mammals. Their deep, interconnected burrows allow many wombats to survive a fire event. This Common Wombat was observed digging under the snow at Lake Mountain, taking advantage of the new green pick that had germinated before winter snowfall.

Sheltering from the fire. Hollows in large trees can shelter animals from fire. The hollow-dwelling Mountain Brushtail Possum, or Bobuck, is a charismatic animal of tall eucalypt forests. A surprising number of Bobucks were able to survive the fire in this way. As part of a long-running research project at Cambarville, east of Marysville, we had fitted radio-transmitting collars to 19 Bobucks in the first week of February 2009. All of these animals survived the fire and were found alive several months afterwards. The broad diet of this species of possum, which includes fungi, fruit, seeds, leaves and insects, enabled it to utilise what little food resources were available immediately after the fire.

The native Bush Rat. The Bush Rat in this cage trap is one of the most abundant mammals in wet eucalypt forests. We have found densities of up to 50 individuals per hectare. The species has distinctive facial features, tail length and colour and can be readily distinguished from introduced rodents that are common in urban areas. Our research has shown that although populations of the Bush Rat are dramatically reduced in severely burnt areas of forest, there are nevertheless small residual populations persisting in the gullies and shallow drainage lines within burnt landscapes.

How do we determine how animal populations recover from the fire? To answer this question, we need to study populations directly after the fire. Rapidly collected information, such as the numbers of animals trapped in burnt areas, as well as genetic information on differences between populations, can reveal whether animals persist in a location or recolonise from unburnt areas. We have found from studies started within months of the 2009 fires that small populations of the Agile Antechinus persist in even the most severely burnt forests. These residual populations help speed the recovery of this species in burnt landscapes. Image: David Lindenmayer

A different survival strategy. (Above) The Swamp or Black Wallaby is highly mobile and known to move through a fire front to seek refuge areas. It can then move out of these green patches to forage on regenerating grasses and fungi in fire-affected areas.

A natural wildlife shelter. (Left) The patchiness of wildfires is critical for the survival of many native animals. Unburnt patches, such as this rainforest gully, provide refuge during the fire and important food resources immediately after the fire. They also are sources of surviving animals that recolonise the surrounding burnt landscape.

The Crimson Rosella was one of the most common bird species detected in our bird surveys soon after the fires. This parrot has been seen in large flocks of up to 30 birds. Like so many aspects of fire ecology, the reasons for these observations are currently unknown.

A winged parasite. Australia has many kinds of cuckoos – birds that lay their eggs in other birds' nests. One of these species is Horsfield's Bronze-Cuckoo, which is a common inhabitant of dry and mixed species foothill forests. After the fires this bird was also recorded in low and moderate severity burnt Mountain Ash forest.

A mobile nomad. The Yellow-faced Honeyeater is a common and widespread bird throughout eastern and south-eastern Australia. Following the fires, it has been recorded in both unburnt and low to medium severity burnt forests.

Logs and rocks provide shelter from fire for some species of reptiles, including the Southern Water Skink. This individual was recorded basking in the forest that was subject to a very high severity burn.

This resprouting Shining Gum is a source of sap for the Yellow-bellied Glider. This tree has had a long history of use by colonies of this gliding marsupial, and has continued to be tapped for its sap after the fire. The bark has been shredded by the glider's teeth, which wounds the tree and causes sap to flow. The Yellow-bellied Glider (above) is unusual among Australian mammals in using eucalypt sap as a substantial component of its diet. Eucalypt sap has some toxic chemicals, including some with properties similar to cyanide.

Tiger Snake. The Tiger Snake is a common predator in unburned Mountain Ash forest, especially in older forests where gaps in the canopies of trees allow sunlight to penetrate to the forest floor. Large numbers of these animals would have perished in the 2009 wildfires, but we have found young animals even in forests subject to very high severity fire. This suggests that residual populations have survived the fire and will be an important part of food webs as the forest recovers. Image: David Lindenmayer

A surprising forest visitor. Australia is the land of parrots with about one-sixth of the world's species occurring here and nowhere else. The Blue-winged Parrot has been a surprising visitor to a number of sites that have begun to recover after the fire.

Taking advantage of the fire. Fires create winners and losers. A big winner is the Flame Robin. Flashes of orange are now common in charred forest. Our repeated field surveys of birds have shown that the Flame Robin occurs in almost every stand of burnt forest, having colonised areas where they were previously long-absent. No-one yet knows how these birds respond so quickly to wildfire and over such large areas.

New arrivals. Some bird species were recorded for the first time in wet eucalypt forests following the 2009 fires. Three of these were the Scarlet Honeyeater, the Red-capped Robin and the White-browed Woodswallow (pictured). Our work over the coming decades will attempt to determine if these recent arrivals are permanent residents or temporary visitors.

Not all species benefit from wildfire. It may take a long time for the forest to again become suitable for some species like the King Parrot.

The ultimate mammalian weed. We have never captured the House Mouse in many years of trapping in undisturbed wet forest. However, this introduced rodent can be common soon after fire in many parts of south-eastern Australia. Our first records of this pest species in the wet forests of Victoria were in previously logged forests that were burnt in 2009.

Some birds are intimately linked to old-growth forests. In Mountain Ash forest, mistletoe occurs only in old-growth stands. As a result, the Mistletoe Bird is also confined to old-growth stands. Therefore, areas of severely burnt old-growth forest may not become suitable for this species for hundreds of years.

A sound of the forest unheard since the fire. The Eastern Whipbird is a common bird in unburnt wet eucalypt forests. This is a large but shy animal, and most visitors to the forest would not have seen it. However, the distinctive whip crack call of the male reveals its presence. This species is strongly reliant on dense ground and shrub cover and we have found it to be virtually absent from forests that have been severely burnt. The way that we survey for the 65 bird species commonly encountered in these forests is by detecting their calls, not by visual observation, due to the height and density of the forest. We complete regular surveys of birds in burnt and unburnt wet forests.

A leaf-eating specialist. Like the Koala, the Greater Glider (above) has a diet composed almost exclusively of eucalypt leaves. Indeed, at about 1 kilogram, the species is almost as small as it is physiologically possible to have a specialised diet of leaves. The Greater Glider is typically absent from burned forests as a result of its strong dependence on eucalypt leaves. However, we have observed the species in places where some green canopy remains and there are nearby suitable nesting hollows (right).

Home is where the hollow is. Many species of animals are dependent on large trees with hollows in wet eucalypt forest. This decay process can take up to 80 years after the death of the tree.

A classic hollow tree, or 'stag'. This tree was killed in the 1939 fires and has remained standing for more than 70 years after its death. Hollows in large-diameter, highly decayed trees such as this one make ideal dens for a number of species, and are particularly favoured by the Mountain Brushtail Possum and Leadbeater's Possum. The wattle surrounding this stag provides an important food source and acts as a sheltered 'bridge' to connect the den tree to the surrounding mid-storey layer. Trees like this are increasingly uncommon in wet eucalypt forest, as they take 300 or more years to develop.

Burning down the house. Fire accelerates the loss of dead standing trees. Those in the late stages of decay are particularly susceptible to collapse after even a low severity fire, such as 'trickle burn'. We found from our long-term surveys of over 160 sites that the 2009 wildfire removed almost all large dead trees in the late stages of decay. This tree stood over 30 metres before the 2009 fires. Over a 10-year history of studying the trees at this site, we recorded the Sugar Glider, the Mountain Brushtail Possum and the Yellow-bellied Glider in this tree before it was burnt.

Trees like the one opposite have been standing for many decades since they were burnt, but are now increasingly susceptible to collapse.

The release of a radio-collared Mountain Brushtail Possum. We conducted a radiotracking study (right) of how these animals use trees with hollows before and after fire (see page 64). Before the fire, these possums preferred dens in trees in the late stages of decay. However, most of these were removed by the fire. The animals responded by switching their choice of den trees to the remaining younger and less decayed standing hollow trees.

Prefabricated homes. Following the fires, some areas of wet eucalypt forest had significant stag loss and will not support trees with hollows for another 60 years. Nest boxes have been established for Leadbeater's Possums in Snow Gums on Lake Mountain. Nest boxes have been made from recycled plastic to prevent them rotting and collapsing.

Spotlighting. We do not normally survey animals by spotlighting in Mountain Ash forests. This is because the understorey is so thick, it is often difficult to see more than a few metres. But after the fires, we were trying to determine which species of possum and glider persisted in burnt landscapes and a few nights of spotlighting were completed. A field diary entry in September 2009 recorded an experience akin to walking through an alien landscape: "*There was not a breath of wind, and it was 24 degrees at 11 pm. Dust from passing salvage logging trucks hung thick in the air like fog … On such a still night in the past you would hear light rustling of the leaves in the canopy, but in this patch of forest, the canopy was completely consumed in the fires, and the only noise I heard was the popping and cracking of the drying Mountain Ash (trees), their tall trunks splitting as they dried out just eight months after they were killed by the fires. Every few seconds there would be another loud crack, either close by or far off in the distance … In five hours of spotlighting, the only living thing I saw were swarms of blowflies which would be stirred up as I walked past, then re-settle as I passed … an uncommon experience in undisturbed forest.*"

The other 99%. Insects are by far the most species-rich animal group in wet forests and comprise a very large proportion of forest biodiversity. However, almost nothing is known about these animals. The study shown above is using small pitfall traps to compare the beetle, ant and spider faunas of burnt and unburnt wet ash forest.

Some common invertebrates of the forest: The huntsman spider (left), bulldog ant (above), Australian wood cockroach (above right, image: Paul Sunnucks), and a nymphal casing of a cicada (below right) that emerged after the fire.

4 | Forest futures

Many people have asked us: Will the forest ever recover? Will the animals come back? As we have outlined in the previous chapters and as is evident in burned areas now, the answer to both is an unequivocal yes. And it is certainly not the first time montane ash forests have recovered from large fires. For researchers and forest managers alike, tracking the recovery process will be a fascinating learning journey over the coming years and decades, and the new knowledge gained will significantly improve the understanding of relationships between wildfire (and natural disturbance in general), forest ecology and biodiversity.

In this chapter, we draw together the knowledge of the fire regime of montane ash forests and briefly examine a number of issues associated with human management practices in forests, forest recovery, fire dynamics and biodiversity conservation. Like the previous chapters, most of our commentary is set in the context of post-fire ecological recovery. However, in this chapter, we also explore the risks posed to forest recovery by altered fire regimes as well as current management practices like logging.

Fire regimes and fire responses

Fire is a critical ecological process in almost all Australian ecosystems, but it is the fire regime that is critical and what has shaped the forests of today. Changes in fire regimes, not individual fire events, will be the biggest natural threat to these forests in the future. Although there are many vegetation types, such as the heathlands and shrublands of coastal eastern Australia or the tropical savannas of northern Australia, that are well adapted to frequent, low severity fires, others, like montane ash forest, are adapted to infrequent, high severity wildfire. Very few ecosystems, however, have evolved to recover from a combination of high frequency and high severity fire.

A series of fires can have either a positive or negative ecological effect and even radically alter entire forest ecosystems. In montane ash forests, the long absence of fires (over 350 years) can lead to eucalypt forests being replaced by cool

Islands of burnt trees. One potential way to mitigate the effects of salvage logging is to retain stands of burnt but uncut trees. The effectiveness of doing this is unknown and an experiment to examine this has therefore recently been established.

temperate rainforest as the mature eucalypts senesce and die. This is because montane ash trees do not regenerate from soil-stored seed, but rather from seeds shed from burned canopy trees (see Chapter 2). Conversely, too much fire also can lead to the loss of the eucalypts in montane ash forest. This is because overstorey eucalypt trees that have not reached maturity will not produce the seed needed for regeneration. Under these circumstances, eucalypt-dominated forest will be replaced by wattle-dominated stands. Some people would view the loss of areas of montane ash forest as problematic and this would be particularly true if it occurred over large areas. However, within landscapes, the presence of areas of cool temperate rainforest and stands of wattle-dominated forest are important vegetation types. They add to the diversity of forest landscapes and are valuable places for some species of plants and animals.

In the specific case of the Central Highlands of Victoria, there is a danger that another severe fire before 2030 will have very different effects on the forest to the 2009 fires. Two fires in relatively quick succession would result in the loss of the eucalypt overstorey from large areas of forest. This loss would have major negative consequences for water production, carbon storage, timber and pulpwood production, and biodiversity conservation. It would be possible to reverse some of these effects by artificial regeneration using seeds gathered from unburned stands, but this would be extremely expensive and involve addressing many important ecological and management considerations, like where to source very large volumes of seed. This problem would be compounded by the fact that areas logged between 2000 and 2030 and then burned in a future wildfire also would need to be re-established through artificial seeding. Furthermore, such seeding would only replace the overstorey (commercial) eucalypt species and not mid-storey and ground cover species.

The impact of climate change

The risks of altered fire regimes are elevated by rapid climate change and associated forecasts of further reductions in rainfall and increases in temperature. Climate change also may have other more subtle, but nevertheless profound, fire-related impacts in montane ash forests. For example, the regeneration niche of montane ash trees may well be altered by rising temperatures. Ash-type tree species may be highly vulnerable to climate change because the regeneration phase is often a critical phase for plant survival.

Although climatic conditions and weather patterns are key drivers of fire behaviour and significantly influence fire regimes, other factors can be important. These include human forest management practices like traditional (green tree) logging and salvage logging. In the following sections we discuss some of the issues surrounding alternative management approaches to montane ash forests. Some of these issues are relevant to the biodiversity of these forests, while others are closely linked to other societal and economic values. How we balance the management of montane ash forests to address different issues depends on the values that we place on these forests.

What do we value in our forests?

Human societies place great value on forests. The montane ash forests of the Central Highlands of Victoria support such diverse values as water production, carbon storage, ecosystem services (pollination, pest control), biodiversity conservation, timber and pulpwood production, tourism, and artistic, spiritual and recreational opportunities. These values are diverse, and differences in the perspectives that individuals hold on these values have generated much controversy. This controversy is created by the challenges associated with a need to manage for the range of key values that native forests provide, and the relative importance different people ascribe to those values. Different stakeholders see some values as primary ones and others as secondary or tertiary values. This may depend on whether they are viewing the forest from a perspective of its potential to provide goods and services, from a purely economic standpoint, or a broader aesthetic, cultural or spiritual perspective. Still others see forests as a fire threat and feel all management should be directed primarily towards reducing this risk. However, debates about forest values are unlikely to ever go away and they are important ones to have in a healthy democracy.

Ecological values

As we have shown in this book, montane ash forests are truly spectacular; the tallest flowering plants in the world are from this region. Some of the world's most enigmatic and iconic animal species are also found in the Central Highlands of Victoria, some of which are threatened or endangered. For example, virtually the entire known distribution of the nationally endangered Leadbeater's Possum occurs in the montane ash forests of this region. This species is also one of the faunal emblems of the State of Victoria. The biodiversity of montane ash forests means they have important tourism, aesthetic and recreational values to the community.

Water production values

Montane ash forests are critically important for the production of most of the water supply for the city of Melbourne. There are important relationships between the age of forests and the amount of water they produce. Immediately after fire or logging, large amounts of run-off occur. However, in subsequent years, vigorously growing young trees transpire large quantities of water through their leaves, resulting in limited additional water available for streamflow and therefore limited water production for human consumption. In contrast, streamflow is highest in old forests where the growth rates of trees have slowed and rates of transpiration are lower than in younger regrowth stands.

Carbon storage values

Montane ash forests have a vital role to play in carbon storage, which is a critical part of sequestering greenhouse gases in efforts to tackle rapid climate change. Old-growth Mountain Ash forests have the highest carbon biomass values of

any kind of forest worldwide – a staggering 2000+ tonnes per hectare. Undoubtedly some carbon is lost from these forests during fire. We are currently repeating measurements for carbon biomass estimates taken before the fires to quantify this. However, what is lost appears to be a relatively small proportion of what was present in the forest before the fire. Even high severity fires typically burn leaves and branchlets, with the main biomass of the forest vegetation (trunks and large branches and majority of underground roots) remaining. These biological legacies carry much of the carbon stored in the pre-fire forest into the new regenerating forest. In fact, the stands of Mountain Ash forest with the highest carbon biomass are old-growth stands where there has been a history of fire, but no salvage logging.

Wood and paper production values

Montane ash forests are important for the production of timber and pulpwood. About 75–80% of the ash-type forest in the region is broadly designated for wood production. In the case of those forests dominated by Mountain Ash, the timber industry is estimated to employ 1000 people in harvesting, log hauling, sawmilling, secondary processing and pulp and paper manufacture. The Victorian Government receives about $1 million annually in royalties from the sale of Mountain Ash sawlogs and about $4 million in royalties per year from the sale of residual logs. Overall, logging in the Mountain Ash forests is estimated to generate around $500 million annually for the Victorian economy. Logging in montane ash forest occurs as clearfell harvesting of unburned forest and salvage logging in burned forest.

Conflicting values

The ways in which we manage montane ash forests for biodiversity conservation, water production, carbon storage, tourism and recreation are largely congruent. and do not involve a change to the historical fire regime of montane ash forests. In contrast, the management of these forests for timber and pulpwood production involves major human disturbances that can have significant tradeoffs for other values. The changes to the forest resulting from logging operations have major impacts on water production, biodiversity conservation and fire regimes. We describe the effects on biodiversity conservation and fire regimes in further detail below.

Traditional logging and wildfire

The traditional kind of logging in montane ash forests is clearfelling. Of the areas of montane ash forest subject to logging, more than 95% are clearfelled. Under this method of cutting, virtually all standing trees are removed from a 15–40 hectare area in a single operation, leaving few live standing trees. Codes of Forest Practice governing the way forests are logged allow for up to three adjacent 40-hectare cut areas. A high intensity 'slash fire' is used to burn logging debris on cut areas (e.g. bark, tree crowns and branches) and to create a nutrient-rich ash seedbed for regeneration of new stands. The planned interval between clearfelling operations on a site is reportedly 80 years, although large areas of forest that were between 40 and 70 years old have been harvested during the past two decades.

Clearfelling has been a highly controversial form of logging in Australian forests, in part because of its impacts on other values like biodiversity conservation. In particular, the method either removes the majority of large old trees from a stand or rapidly accelerates the decay and collapse of those large trees which are retained. These key attributes of forests do not redevelop in a harvested area for a century or more. The loss of such forest 'infrastructure' is a major problem for species that depend on large old trees, like the array of animals dependent on tree hollows for sheltering and nesting. Our recent research has attempted to develop alternative logging methods to clearfelling that better retain the key features of forests needed by biodiversity. One that shows some promise involves the retention of islands of forest within cut blocks of forest. This approach attempts to mimic some of the natural patchiness of burned and unburned areas that can be created by wildfires and other disturbance – a pattern to which some elements of the biota appear to be well adapted. A key practical aim of the approach is to allow remnants of older forest to be retained within surrounding young, post-logging regeneration. This creates the combination of old trees and regrowth forest or 'multi-aged' forest that is more likely to provide suitable habitat for a number of key species in montane ash forest such as the endangered Leadbeater's Possum.

Although the effects of clearfelling on biodiversity have been reasonably well established for several decades, relationships between logging and wildfire have only begun to emerge in the past 10–20 years. Research in rainforests and other kinds of very wet forest around the world suggests that logging makes them more fire-prone. This is because logging changes the structure and composition of forests in ways that can have a major influence on fires, including:

1. **Changing microclimates and fuel characteristics.** These changes can lead to increased drying of understorey vegetation and the forest floor, which can alter the amount, type and moisture content of fuels.

2. **Altering the structure and plant species composition of stands.** Such changes not only alter microclimatic conditions as described above, but also can change densities and patterns of trees, inter-crown spacing of trees and other forest attributes like plant species composition. This can, in turn, influence fire behaviour and eventually even change fire regimes. For example, clearfelling of wet forests in southern Australia leads to the development of dense stands of regrowth saplings which can create more available fuel than if the forest is not clearfelled.

3. **Altering patterns of landscape cover.** Logging operations change natural patterns of spatial juxtaposition of different kinds of forests stands. This, in turn, may change the way fire spreads through landscapes. For example, some areas traditionally characterised by an absence of fire may become more susceptible to being burned by fires that spread from adjacent, more flammable, logged areas.

In summary, although forests are regenerated after logging, many characteristics of such regrowth forests may make them more – not less – prone to some aspects of fire regimes such as increased fire severity. This is an important new perspective on the inter-relationships between fire management and forest management which requires careful thought, especially relative to the location of logged areas and human settlements.

Salvage logging

Wildfire is a messy process – it leaves huge quantities of ash, large numbers of dead trees and burned logs, and fire-scarred living trees. These are the biological legacies of the previous stand (see Chapter 2). Post-fire salvage logging is common after major disturbances like the 2009 wildfires and it aims to recover some of the economic value of fire-damaged trees. Ecologically, however, dead trees created by a fire are important and valuable biological legacies (see Chapter 2).

Clearfelling is the typical form of salvage logging in montane ash forests and it is therefore a double set of disturbances – wildfire followed by logging. In some cases there may be a third disturbance as additional management efforts are needed to regenerate salvage logged areas.

Salvage logging has a range of ecological impacts on biodiversity. Three key ones are:

- The removal of biological legacies, impairing the many important ecological roles they play (see Chapter 2).
- A simplification of forest structure. Salvage logging removes most of the standing timber, and the forest which regenerates is a single age class. In contrast, multi-aged montane ash forests are very important environments for a range of species. For example, the highest species diversity of arboreal marsupials occurs in multi-aged forests.
- The reduced survival of some native plant species, particularly those that have been stimulated to germinate or resprout by a wildfire but which may be killed or damaged by logging machinery before they have a chance to mature and set more seed.

The past impacts of salvage logging in montane ash forests, particularly those from the prolonged salvage logging operations after the 1939 wildfires, may have been far more substantial than appreciated by many forest managers and conservation biologists. For example, the removal of considerable numbers of large fire-damaged trees and the loss of many areas which would have been multi-aged forest has undoubtedly contributed to the shortage of trees with hollows that now characterises substantial areas of montane ash forest in the Central Highlands of Victoria. In an effort to better understand the impacts of salvage logging and develop ways to mitigate those impacts, we have recently commenced a new experiment involving the retention of islands of burnt forest within salvage logging coupes. It is designed to compare the recovery of the flora and fauna within retained islands with ecological recovery in traditional salvage logged areas.

There may be other environmental impacts of salvage logging beyond those on biodiversity. First, the removal of living and dead trees reduces the amount of carbon that occurs in a stand. A burnt forest that is not salvage logged will

develop a complex structure composed of young regenerating trees plus living and/or dead trees remaining from the previous stand. Large amounts of carbon biomass can be stored in montane ash forests in this way – significantly more than in burned and subsequently salvage logged areas. Maintaining high levels of carbon storage in forests is a crucial part of sequestering greenhouse gases and tackling climate change. Second, mechanical disturbance of soil leads to a further loss of soil and nutrients from erosion and compaction. Up to 85% of the soil in a salvage logged area can be disturbed. These already fragile soils are at a crucial stage of natural recovery following fire. Such secondary disturbances can have serious negative effects on an ecosystem which has already begun the first stages of recovery – the first flush of seedling regeneration can be lost and other nutrients, seeds and spores found in the soil will be affected.

Concluding comments

This book has focused on the process of post-fire ecological recovery with the aim of describing its complexity but also trying to simplify the science in a way that makes that complexity understandable. The array of fire response mechanisms employed by plants and animals means the recovery process after the 2009 wildfires is already well established. But for some species and some characteristics of a natural forest, recovery will be slow. We will have to accept that much of the burned forest will not be the same again within our lifetimes. However, it is important to understand that it **is** recovering, and to be patient observers of nature during this process. This is despite the fact that people in 'developed' societies are generally not good at either being patient or observing nature. Nor are they good at comprehending the time dimension of forests – some of the trees burned on 7 February 2009 preceded the arrival of Captain Cook by more than a century and the seedlings that have just replaced them may not rival their predecessors' age until the middle of this millennium.

Should montane ash forests be prescribed burned?

A recurrent theme after almost all major conflagrations in Australia is a public outcry that insufficient fuel reduction/prescribed burning was done prior to 'the wildfires' and that if it had been done then 'none of this would have happened'. In the case of montane ash forests, these are unfortunate comments generally made in the absence of ecological understanding. Any burning of a young regenerating forest comprising trees less than 20–30 years old risks killing all of the trees and losing the forest altogether (along with the myriad of plant and animal species associated with it). Even within older montane ash forests, prescribed burning is an ecologically inappropriate management strategy. This is because montane ash forests are generally very wet and burn only under extreme conditions when the only kind of fire that will occur is a wildfire that is either stand-replacing or partially stand-replacing.

Learning to read the landscape. The long history of wildfires in wet eucalypt forests can be seen across the ridges and valleys of this landscape. The dead standing trees in the upper left of the image are legacies of an old-growth forest burnt in 1939, and provide shelter for many animals. The gully running from the centre to the top right supports cool temperate rainforest, which would have been an important refuge for animals and plants to recolonise the burnt landscape after the 1939 fire. The fire of February 2009 approached from the upper left of the image and burnt the forest canopy on the north-western flank of the ridge in the foreground. A new stand of Mountain Ash trees will regenerate from the seed falling from this scorched canopy in that area. The 70-year-old forest on the south-eastern flank of this ridge would have regenerated following 1939, but escaped the fire in 2009 probably due to different wind directions. If not burnt or logged, it will become old-growth forest over the next 100 to 200 years. This image shows how the wildfires shape changes in a forest landscape.

An ancient forest, burnt but standing. Our research site #470 was an old-growth stand deep in the closed O'Shannassy water catchment. This is the first major high severity fire in this stand for at least 250 years. These Mountain Ash trees, over 70 metres tall, have not survived the 2009 fires and will soon be surrounded by a new generation of young trees. However, the huge Shining Gum in the far left background is rapidly recovering by epicormic sprouting. We began monitoring this site in 1987. The large number of hollow-bearing trees supported many species of hollow-dwelling animals over the past decades, including the Mountain Brushtail Possum, the Sugar Glider, the Greater Glider, Leadbeater's Possum and the Yellow-bellied Glider. While only one of these species has been detected at this site in the year after the fire (Mountain Brushtail Possum), the legacies of the old-growth forest will eventually provide shelter for many of them as the stand regenerates over coming decades.

Different disturbances, different impacts. This set of images shows how burnt and logged forest recovers from a very different starting point to burnt but unlogged forest. Compared to the unburnt forest on the lower slopes of this landscape, the burnt and unlogged forest has lost its understorey and mid-storey vegetation. Despite the loss of the old, decayed dead standing trees, most of the burnt trees are still standing, providing shelter for wildlife. The scorched canopy has dropped its leaves and seeds onto the ground as part of beginning the process of forest regeneration. The forest on the ridge top was logged before the 2009 fires. Very few hollow trees characterise this forest patch and there is little eucalypt regeneration at ground level due to the prior absence of a seed-bearing canopy.

Where are the eucalypts? This forest stand has been burnt twice within the past 20–30 years. Insufficient time had elapsed between the two fires for the eucalypts to reach maturity and produce seed. Therefore, this stand is now dominated by wattles. A similar kind of forest develops where a recently logged and regenerating young forest is burnt in a fire.

A naturally patchy wildfire. Unburned patches within the boundaries of a fire, such as the Myrtle Beech dominated rainforest in this gully, are critically important refuges for the persistence of many species in burnt landscapes. These areas should be excluded from logging and backburning.

Violin voice. The call of the Rufous Fantail is sometimes said to resemble the descending crescendo of a poorly played group of violins. We have not heard such 'music' in the forests severely burnt by the fires. However, we have found the Rufous Fantail surviving in unburnt refuges within the fire-affected area. Its recovery will be dependent on survival in these refuges before it can recolonise the regenerating burnt forest.

What happens before a fire matters. This forest stand was logged before the fires in 2009. Wet eucalypt forests are highly sought after for the production of timber, pulpwood and woodchips. Fire and logging are very different disturbances to this ecosystem, and the pattern of post-fire recovery of a logged forest is very different to that of an unlogged one. The old-growth understorey species, such as tree ferns and the Musk Daisy Bush, are usually badly damaged or killed by logging. Very few structural legacies are retained and seed fall can be highly patchy or absent depending on how the tree canopies are spread or heaped for burning post-harvesting.

An island experiment. Part of our work in montane ash forests has involved experimenting with new kinds of logging that are alternatives to clearfelling: the retention of islands in cut forest supports populations of species that would otherwise be absent from recently clearfelled coupes. The large trees retained in these islands will provide important shelter resources (such as hollows) for animals in these landscapes as the young forest regenerates after logging.

Salvage logging. This is a typical management response to high severity wildfires in wet eucalypt forest. Fire-damaged trees are removed from extensive areas of burnt landscapes. Salvage logging takes place quickly after fires, before the timber deteriorates and is unsaleable. Salvage logging is a double disturbance, as a fire is followed soon after by intensive clearfell logging including mechanical disturbance from harvesting machinery. Up to 85% of the soil layer is permitted to be disturbed in this process. An ecological consequence is that germinating plants across much of the harvested area are killed. Furthermore, there is a major loss of biological legacies, such as the logs and standing dead trees that are critical to the recovery of populations of animals like Leadbeater's Possum. The impacts of salvage logging on the forest can be prolonged, possibly for as long as 200 years.

Salvage logging and fire risk. Work in the wet coniferous forests of western Oregon (USA), which are an overseas analogue of Mountain Ash forest, has suggested that the fine fuels and additional debris created by salvage logging can increase the short- and medium-term risk of fire. There has been no equivalent work in Australian forests but it is a knowledge gap that demands urgent attention.

Fire-prone forests? Managing the risk of fire in wet eucalypt forests is a controversial issue. Some people assume that logging a forest makes it 'safer'. However, research in Victoria and overseas suggests that young, densely regenerating forest may be more prone to fire, and burns at a higher severity than unlogged forest.

An endangered plant recovering. The biodiversity values of wet eucalypt forests are immense, even after a severe wildfire. This extremely rare plant, the Shiny Nematolepis, was known from only a single wild population in the O'Shannassy water catchment. All of the adult plants of this species were killed by the fire in February 2009. Botanists were unsure whether the species would regenerate adequately from seed. However, surveys a year after the fires have shown a healthy regenerating population of several thousand seedlings. While this population is recovering from the fire, the species will be highly susceptible to a second disturbance before the plants reach maturity. A second population has since been discovered in another area and a third population has been planted in yet another location to try to minimise the risks associated with such limited distribution.

An endangered animal recovering. Fire in old-growth eucalypt forest produces habitat suitable for the Leadbeater's Possum within 10–15 years. This is because the combination of dense regenerating forest and old hollow trees from the previous stand provides important shelter and connectivity for the animals to move through the forest. However, these conditions do not occur after the burning of young forest because the trees are not large enough to develop suitable cavities for nesting. Legacies of old-growth forest that was burnt in the 1983 Ash Wednesday fires have provided excellent habitat for Leadbeater's Possum. In contrast, we have not found these rare possums in stands where the pre-fire legacies were not retained after the Ash Wednesday fires. This demonstrates that changes to logging practices can have a significant positive impact on one of Victoria's faunal emblems.
Image: Mike Greer and David Lindenmayer

History repeats itself. Fire is an important and integral part of forest ecosystems. Montane ash forests have recovered repeatedly after fire over hundreds of thousands of years. They will do so again – particularly if they are carefully managed and conserved.

A massive store of carbon. Montane ash forests store massive amounts of carbon. Values of 2000 tonnes of above-ground biomass have been recorded from old-growth stands, making these the most carbon-dense forest stands in the world. Notably, although some of the most carbon-dense stands were killed by the 2009 wildfires, the majority of the carbon in these areas remains – held within dead trees and fallen logs. Over time, the combination of these dead biological legacies and dense regenerating new forest will create a spike in the amount of carbon that is stored. Image (top): Esther Beaton

Water for the people. Water production is a significant 'service' to humans provided by wet eucalypt forests. Most of Melbourne's population currently depends on water derived from montane ash forests. Fire and logging change the age and structure of the forest, which affects the amount of water transpired by regenerating trees and surface run-off patterns. Consequently, stream flow levels and water yields are reduced during the forest recovery phase. The long-term integrity of these water catchments is therefore intimately tied to the long-term integrity of the city of Melbourne.

LIST OF SCIENTIFIC NAMES

PLANTS

Eucalypts

Eucalyptus baxteri	Brown Stringybark
Eucalyptus cypellocarpa	Mountain Grey Gum
Eucalyptus dalrympleana	Mountain Gum
Eucalyptus delegatensis	Alpine Ash
Eucalyptus dives	Broad-leaf Peppermint
Eucalyptus nitens	Shining Gum
Eucalyptus obliqua	Messmate
Eucalyptus pauciflora	Snow Gum
Eucalytus radiata	Narrow-leaf Peppermint
Eucalyptus regnans	Mountain Ash
Eucalyptus ovata	Swamp Gum
Eucalyptus goniocalyx	Long-leaved Box
Eucalyptus macrorhyncha	Red Stringybark

Mid-storey trees

Acacia dealbata ssp. *dealbata*	Silver Wattle
Acacia melanoxylon	Blackwood
Acacia obliquinervia	Mountain Hickory Wattle
Atherosperma moschatum	Southern Sassafras
Lomatia fraseri	Tree or Forest Lomatia
Nothofagus cunninghamii	Myrtle Beech
Pomaderris aspera	Hazel Pomaderris

Mid-storey shrubs

Correa lawrenceana var. *latrobeana*	Mountain Correa
Nematolepis wilsonii (syn *Phebalium wilsonii*)	Shiny Nematolepis
Olearia argophylla	Musk Daisy-bush
Pimelea axiflora ssp. *axiflora*	Bootlace Bush

Lower storey: herbs and groundcovers (< 1 m)

Acaena novae-zelandiae	Bidgee-widgee
Senecio glomeratus ssp. *glomeratus*	Annual Fireweed
Senecio gunnii	Mountain Fireweed
Senecio minimus	Shrubby Fireweed
Senecio vagus ssp. *vagus*	Saw Groundsel
Senecio velleioides	Forest Groundsel
Chiloglottis valida	Common bird orchid

Ferns and fern allies

Calochlaena dubia	Common Ground-fern, False Bracken
Pteridium esculentum	Austral Bracken
Cyathea australis	Rough Tree-fern
Dicksonia antarctica	Soft Tree-fern

Weeds

Rubus fructicosus	Common Blackberry
Pinus radiata	Monterey Pine, Radiata Pine
Cirsium vulgare	Spear Thistle
Sonchus oleraceus	Sow Thistle, Milk Thistle
Hedera helix	Ivy
Ilex aquifolium	Holly

FUNGI

Cyttaria gunnii	Beech Orange
Tremella mesenterica	
Mycena austrororida	
Pycnoporus coccineus	
Neurospora sp.	

ANIMALS

Mammals (placental)

Mus musculus	House Mouse
Sambar unicolor	Sambar Deer
Rattus fuscipes	Bush Rat

Mammals (marsupial)

Vombatus ursinus	Common Wombat
Wallabia unicolor	Swamp Wallaby, Black Wallaby
Antechinus agilis	Agile antechinus
Trichosurus cunninghami	Mountain Brushtail Possum, Bobuck
Petaurus australis	Yellow-Bellied Glider
Petauroides volans	Greater Glider
Gymnobelidus leadbeateri	Leadbeater's Possum
Phascolarctos cinereus	Koala

Reptiles

Eulamprus tympanum	Southern Water Skink, Highlands Water Skink
Notechis scutatus	Tiger Snake

Birds

Petroica phoenicea	Flame Robin
Calyptorhynchus funereus	Yellow-tailed Black Cockatoo
Platycercus elegans	Crimson Rosella
Tyto tenebricosa	Sooty Owl
Eopsaltria australis	Eastern Yellow Robin
Neophema chrysostoma	Blue-winged Parrot
Myzomela sanguinolenta	Scarlet Honeyeater
Lichenostomus chrysops	Yellow-faced Honeyeater
Petroica goodenovii	Red-capped Robin
Artamus superciliosus	White-browed Woodswallow
Alisterus scapularis	King Parrot
Psophodes olivaceus	Eastern Whipbird
Dicaeum hirundinaceum	Mistletoe Bird
Chrysococcyx basalis	Horsfeld's Bronze-Cuckoo

Invertebrates

Panesthia australis	Australian Wood Cockroach

FURTHER READING

Ashton DH (1975) The root and shoot development of *Eucalyptus regnans* F. Muell. *Australian Journal of Botany* **23**, 867–87.

Ashton DH (1981) Tall open forests. In *Australian Vegetation*. (Ed. RH Groves), pp. 121–51. Cambridge University Press, Cambridge.

Banks SC, Dujardin M, McBurney L, Blair D and Lindenmayer DB (2010) Starting points for small mammal population recovery after wildfire: recolonization, refugia or residual populations? *Oikos* (in press).

Boland DJ, Brooker MI, Chippendale GM, Hall N, Hyland BP, Johnston RD, Kleinig DA, McDonald MW and Turner JD (2006) *Forest Trees of Australia*. 5th edn. CSIRO Publishing, Melbourne.

Bradstock RA, Williams JE and Gill AM (Eds) (2002) *Flammable Australia: The Fire Regimes and Biodiversity of a Continent*. Cambridge University Press, Melbourne.

Flint A and Fagg P (2007) *Mountain Ash in Victoria's State Forests. Silviculture Reference Manual No. 1*. Department of Sustainability and Environment, Melbourne.

Kraaijeveld-Smit FJ, Lindenmayer DB, Taylor AC, MacGregor C and Wertheim B (2007) Comparative genetic structure reflects underlying life histories of three sympatric small mammal species in continuous forest of south-eastern Australia. *Oikos* **116**, 1819–30.

Lindenmayer DB (1996) *Wildlife and Woodchips: Leadbeater's Possum as a Testcase of Sustainable Forestry*. New South Wales University Press, Sydney.

Lindenmayer DB (2009a) *Forest Pattern and Ecological Process: A Synthesis of 25 Years of Research*. CSIRO Publishing, Melbourne.

Lindenmayer DB (2009b) Old forests, new perspectives: insights from the wet ash forests of the Central Highlands of Victoria. *Forest Ecology and Management* **258**, 357–65.

Lindenmayer DB and Beaton E (2000) *Life in the Tall Eucalypt Forests*. New Holland Publishers, Sydney.

Lindenmayer DB, Burton PJ and Franklin JF (2008) *Salvage Logging and Its Ecological Consequences*. Island Press, Washington DC.

Lindenmayer DB, Hunter M, Barton P and Gibbons P (2009) The effects of logging on fire in moist forests. *Conservation Letters* (in press).

Lindenmayer DB and Ough K (2006) Salvage harvesting in the montane ash forests of the Central Highlands of Victoria, south-eastern Australia. *Conservation Biology* **20**, 1005–15.

Mackey B, Lindenmayer DB, Gill AM, McCarthy MA and Lindesay JA (2002) *Wildlife, Fire and Future Climate: A Forest Ecosystem Analysis*. CSIRO Publishing, Melbourne.

Noble WS (1977) *Ordeal by Fire. The Week a State Burned Up*. Hawthorn Press, Melbourne.

Steffen W, Burbidge A, Hughes L, Lindenmayer DB, Musgrave W, Stafford-Smith M and Werner P (2009) *Australia's Biodiversity and Climate Change*. CSIRO Publishing, Melbourne.